Collins

OCR GCSE 9-1
Computer Science

Workbook

Paul Clowrey

Preparing for the GCSE Exam

Revision That Really Works

Experts have found that there are two techniques that help you to retain and recall information and consistently produce better results in exams compared to other revision techniques.

It really isn't rocket science either – you simply need to:

- **test yourself** on each topic as many times as possible
- **leave a gap** between the test sessions.

Three Essential Revision Tips

1. Use Your Time Wisely

- Allow yourself plenty of time.
- Try to start revising six months before your exams – it's more effective and less stressful.
- Don't waste time re-reading the same information over and over again – it's not effective!

2. Make a Plan

- Identify all the topics you need to revise.
- Plan at least five sessions for each topic.
- One hour should be ample time to test yourself on the key ideas for a topic.
- Spread out the practice sessions for each topic – the optimum time to leave between each session is about one month but, if this isn't possible, just make the gaps as big as realistically possible.

3. Test Yourself

- Methods for testing yourself include: quizzes, practice questions, flashcards, past papers, explaining a topic to someone else.
- Don't worry if you get an answer wrong – provided you check what the correct answer is, you are more likely to get the same or similar questions right in future!

Visit **collins.co.uk/collinsGCSErevision** for more information about the benefits of these revision techniques and for further guidance on how to plan ahead and make them work for you.

Command Words used in Exam Questions

This table defines some of the most commonly used command words in GCSE exam questions.

Analyse	Focus on, and when possible break down, the key elements that can be used to build conclusions.
Complete	Provide any missing information or visual elements.
Convert	Change data from its original form to another.
Define	Precisely provide the meaning of a term, phrase or attribute.
Describe	Provide in your own words an account of a specific term, item or process.
Design	Create your own response to meet a defined set of requirements.
Explain	Describe with additional detail that supports and gives reasons for your answer.
How	In what way, or by what means, did something happen?
Identify	Make an informed choice from a range of possible answers.
Label	Add key terms or short descriptions to the diagram or graphic provided.
List	Provide one or more specific answers without elaboration.
State	Provide, without elaboration, a specific answer or example.

Contents

* Grading information is for guidance only and is not endorsed by the examination board.

Systems Architecture, Memory and Storage

The Purpose and Function of the Central Processing Unit

1 Match each term with its definition. [3]

Input		Programmed instructions for a computer with a specific task.
Output		A device that sends information or data to the CPU.
Software		Any physical device that would normally form part of a computer system.
Hardware		A device that receives instructions or data from the CPU.

2 Complete the representation of von Neumann architecture using the letters from the box. [3]

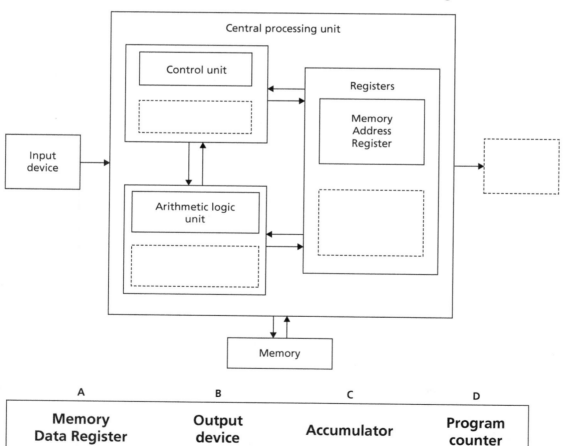

	A	B	C	D
	Memory Data Register	Output device	Accumulator	Program counter

GCSE Computer Science

Systems Architecture, Memory and Storage

Video Solution   Question 2

Grade 4–6

3 Insert the missing word to complete the following sentence.

Registers either store _____ locations or the actual data to be processed. **[1]**

Grade 4–6

4 In von Neumann architecture, what is meant by the term 'stored program'? **[2]**

_____

_____

Grade 7–9

5 Describe the relationship between the ALU and the accumulator. **[2]**

_____

_____

Grade 7–9

6 In von Neumann architecture, what part of the CPU is the program counter found in? **[1]**

_____

Systems Architecture

Grade 1–3

1 Define the term 'clock speed'. **[1]**

_____

Grade 1–3

2 A modern washing machine has embedded systems.

Describe **two** benefits that this might offer the manufacturer. **[2]**

_____

_____

_____

Grade 4–6

3 The clock speed of modern CPUs has slowed in recent years. Instead, CPUs now have multiple cores.

Explain how this change has affected CPU performance. **[2]**

_____

_____

Grade 4–6

4 Briefly describe **four** stages of the fetch–decode–execute cycle. [4]

....

....

....

....

Grade 7–9

5 What is the difference between the L1 cache and the L2 cache? [2]

....

....

Grade 7–9

6 A computer gamer is finding that she has issues with her new console. Some games are crashing and it shuts down for no reason. She has been told it might be caused by one of the embedded systems.

Describe **two** disadvantages of multiple embedded systems in a device. [2]

....

....

....

Memory

Grade 1–3

1 In a modern computer system, what is the purpose of RAM? [1]

....

Grade 1–3

2 RAM is described as volatile, whereas ROM is described as non-volatile.

a) What do the terms 'volatile' and 'non-volatile' mean? [2]

....

....

b) Why is this difference important to computer manufacturers when they are designing BIOS? [1]

....

Systems Architecture, Memory and Storage

Grade 1–3 **3** Define the term 'primary storage'. **[1]**

....

Grade 1–3 **4** Primary storage is normally found on the motherboard.

State whether this statement is true or false. **[1]**

....

Grade 4–6 **5** When a computer is booted, the BIOS runs.

a) What does the abbreviation BIOS stand for? **[1]**

....

b) State two tasks the BIOS carries out as the computer starts. **[2]**

....

....

Grade 4–6 **6** Virtual memory is created by the operating system when required.

a) Why is virtual memory created? **[1]**

....

b) Where is virtual memory created? **[1]**

....

c) What problem can virtual memory cause? **[1]**

....

Systems Architecture, Memory and Storage

Storage Types, Devices and Characteristics

Grade 1–3

1 Describe **two** characteristics of magnetic storage that prevent it from being used in mobile devices such as smartphones and action cameras. **[2]**

....

....

Grade 4–6

2 A teacher is concerned about losing work. She is looking for a secondary storage device that will back up important files.

List **five** characteristics the teacher should consider when choosing a secondary storage device. **[5]**

....

....

....

....

....

Grade 4–6

3 Optical discs are a popular and cheap form of storage.

a) Name **three** different variations of optical disc. **[3]**

....

....

....

....

b) State **three** issues that should be considered when using optical discs. **[3]**

....

....

....

....

Systems Architecture, Memory and Storage

4 After many years of using his magnetic storage-based 120 GB MP3 player, Frank decides to upgrade to a 64 GB SSD-based model.

Describe **one** advantage that each device has over the other. **[2]**

....

....

....

5 Why would an optical drive not be suitable as the main storage for a laptop? **[2]**

....

....

....

6 Technology experts often warn users to think about the lifespan of storage media when considering file backup storage solutions.

Describe **one** way a user can prevent their current backup storage solution becoming outdated. **[1]**

....

....

Systems Architecture, Memory and Storage

Units and Formats of Data

Grade 4–6

1 Match each typical file type with its associated unit of data. **[3]**

high-definition video	terabytes
MP3 audio file	kilobytes
system backup files	megabytes
word processing document	gigabytes

Grade 4–6

2 What are the 8-bit ASCII numbers of the following characters?

   a) The UK pound symbol **[1]**

_____

   b) The division symbol **[1]**

_____

   c) An upper-case K **[1]**

_____

   d) An exclamation mark **[1]**

_____

Grade 7–9

3 What are 'control characters', 'printable characters' and 'symbols and punctuations' examples of? **[1]**

_____

Grade 7–9

4 What is the name of the character set designed as a world industry standard? **[1]**

_____

Grade 7–9

5 Why was an extra zero added to the original version of 7-bit ASCII? **[2]**

_____

_____

Systems Architecture, Memory and Storage

Grade 7–9

6 State **three** non-printed commands that can be represented using a character set such as ASCII. **[3]**

_____

_____

_____

Converting Data 1

Grade 1–3

1 Convert the following denary numbers into binary.

a) 18 **[1]**

_____

b) 25 **[1]**

_____

c) 251 **[1]**

_____

d) 161 **[1]**

_____

Grade 4–6

2 Add the following binary numbers and then convert the answer into denary.

a) 00000101 + 11100101 **[2]**

_____

_____

b) 00110101 + 00100100 **[2]**

_____

_____

Systems Architecture, Memory and Storage

Grade 4–6

3 Write a binary addition calculation that will result in a binary overflow.   **[1]**

....

Grade 7–9

4 Carry out a left shift of 1 on the following binary numbers.

   a) 00110101   **[1]**

....

   b) 00100110   **[1]**

....

Grade 7–9

5 Carry out a right shift of 1 on the following binary numbers.

   a) 11010100   **[1]**

....

   b) 01011010   **[1]**

....

Grade 7–9

6 What is the practical purpose of using left and right shifts with binary numbers?   **[1]**

....

Converting Data 2

Grade 4–6

1 Convert the following denary numbers into hexadecimal.

   a) 254   **[1]**

....

   b) 99   **[1]**

....

   c) 42   **[1]**

....

Systems Architecture, Memory and Storage

2 Using the ASCII table, the word BIG can be represented in binary as:

01000010   **01001001**   **01000111**

Convert this binary sequence into a hexadecimal number sequence.   **[3]**

3 Convert the following hexadecimal numbers into denary.

    a) 4F   **[1]**

    b) 8C   **[1]**

    c) 12   **[1]**

4 Why might a programmer claim that hexadecimal is easier to work with than binary?   **[1]**

5 What mistake might a new programmer make when working with the hexadecimal number 24?   **[1]**

6 If a computer can understand the binary number system, explain why it cannot understand hexadecimal.   **[2]**

Audio/Visual Formats and Compression

Grade 4–6

1 How many bits would be needed to create an image with:

a) 2 colours? [1]

_____

b) 4 colours? [1]

_____

c) 16 colours? [1]

_____

Grade 4–6

2 State **two** ways in which a photo management program will use metadata. [2]

_____

_____

Grade 4–6

3 Why is the RAW image format popular with professional photographers? [1]

_____

Grade 4–6

4 Audio podcasts are very popular with users of smartphones.

What file type is normally used, and why? [3]

_____

_____

_____

Grade 7–9

5 Describe a benefit and a drawback of sampling an analogue audio recording at a very high bit depth. [2]

_____

_____

Grade 7–9

6 Why might using JPEG files cause a problem for a designer who regularly edits the same images more than once? [2]

_____

_____

Computer Networking

> ## Wired and Wireless Networks 1

Grade 1–3

1 Describe what the abbreviations LAN and WAN stand for, and the relationship between them.

[4]

....

....

....

....

Grade 1–3

2 The Internet is described as the world's largest WAN, connecting LANs across the world.

State **two** connection methods for information to travel between countries. [2]

....

....

Grade 1–3

3 Circle which of the following modern devices commonly include network connectivity. [6]

Toaster   Smart TV   Smartphone   Vacuum cleaner

Internet radio   Games console   Tablet   Microwave

Media streamer   Audio amplifier   Desk lamp

Grade 4–6

4 Grace's Gadgets has a new office with a large LAN, but staff are puzzled as to why their network is so slow.

Describe **three** possible factors they should consider. [3]

....

....

....

Grade 7–9

5 Tariq has been using a peer-to-peer network for a while, but he is now concerned that some files are missing from his computer.

Why might this be the case? [2]

....

....

Computer Networking

Grade 7–9

6 Describe **two** advantages and **two** disadvantages of a client–server network. **[4]**

....

....

....

....

Wired and Wireless Networks 2

Grade 1–3

1 a) Describe **three** ways in which cloud computing would be beneficial to an international travel reporter. **[3]**

....

....

....

b) Using cloud technology, the travel reporter does find **one** problem while working away from home. What is it? **[1]**

....

Grade 4–6

2 Describe the difference between a switch and a router. **[2]**

....

....

Grade 4–6

3 Put the following transmission methods in order of potential distance of data travelled, with the shortest first. **[2]**

Fibre optic, Ethernet, Wi-Fi

....

Grade 4–6

4 If the Internet we know and use today is a vast interconnected collection of networks, then what is the World Wide Web? **[2]**

....

....

Computer Networking

Grade 7–9

5 What is a DNS and how does it help users browse the Internet? **[2]**

....

....

Network Topologies

Grade 4–6

1 Draw diagrams to represent a star network topology and a mesh network topology. **[2]**

Grade 4–6

2 A nursery school wants to create a small network but cannot decide between a star and a mesh topology.

Provide **one** key advantage and **one** key disadvantage of each topology. **[4]**

....

....

....

....

....

Computer Networking

Grade 4–6 **3** What device, which is part of most network topologies, will cause the network to fail if it is damaged, and why? **[2]**

_____

_____

Grade 4–6 **4** Why does a mesh network require more cabling than other topologies? **[1]**

_____

Grade 7–9 **5** Which network topology will not slow down by increasing the number of connected devices? **[1]**

_____

Protocols and Layers

Grade 1–3 **1** What does Gbps stand for? **[1]**

_____

Grade 1–3 **2** **a)** While visiting a fast-food restaurant, Jack receives a smartphone notification warning him about connecting to an open unsecure network.

Why should he be concerned? **[3]**

_____

_____

_____

b) What should Jack look for when connecting to Wi-Fi networks? **[2]**

_____

_____

Computer Networking

3 An elderly couple setting up their first home computer and a mobile florist business have both set up their first email accounts.

Which email protocol might be the most appropriate for each, and why? **[4]**

....

....

....

....

4 Which protocol layer is primarily concerned with the communication of IP addresses between network routers? **[1]**

....

5 Describe how hardware and software standards might be applied to the design of a new laptop computer. **[4]**

....

....

....

....

....

System Security and Software

Common System Threats

Grade 1–3

1 Online shopping sites and social networks are often the focus of online attacks designed to steal personal information.

State **four** pieces of information that cyber-criminals would consider valuable. **[4]**

....

....

....

....

Grade 1–3

2 Match each term with its definition. **[2]**

Pharming		Installed by a user who thinks it is a legitimate piece of software when it is, in fact, malware.
Ransomware		The redirection from a user's website to a fraudulent site, by modifying their DNS entries.
Trojan		Limits or denies a user access to their system until a ransom is paid.

Grade 4–6

3 An anti-malware scan on an infected computer lists the presence of both a worm and a virus.

What is the difference between a worm and a virus? **[2]**

....

....

Grade 4–6

4 Describe how a phishing scam might allow a cyber-criminal to gain someone's bank details via telephone. **[3]**

....

....

....

Grade 7–9

5 A controversial pop star has had their website closed by a DoS attack, and has been told that this will keep happening until a song is withdrawn.

Describe how this might have happened. **[3]**

....

....

....

Grade 7–9

6 A network policy informs members of an organisation about what they are and are not allowed to do when they are using on-site computers.

State **three** aspects of a user's work life that might be covered by this policy. **[3]**

....

....

....

Threat Prevention

Grade 1–3

1 Why must anti-malware be updated regularly? **[2]**

....

....

Grade 1–3

2 A social network would like users to improve their security settings, and it wants to provide a list of five top tips for creating a strong password.

State **five** pieces of advice the social network might offer in its list. **[5]**

....

....

....

....

....

System Security and Software

Grade 4–6

3 Describe **three** ways in which encryption can improve the security of an organisation's network. [3]

....

....

....

Grade 4–6

4 A hospital has been advised to set user access levels for the staff who use its computer network.

State **three** possible reasons for this advice. [3]

....

....

....

Grade 7–9

5 An organisation that is concerned about its online security hires an expert in penetration testing to try to access its system.

What will the expert be looking for during the penetration testing? [3]

....

....

....

Grade 7–9

6 Describe the difference between a public encryption key and a private encryption key. [2]

....

....

....

System Security and Software

System Software

Grade 1–3

1 Describe **two** user benefits of a graphical user interface. **[2]**

....

....

....

Grade 1–3

2 Match each term with its definition. **[3]**

Third-party applications	The process of reducing the file size of a computer file to use less disk space.
Utility software	The software link between the hardware, software and user.
Operating system	Performs specific tasks, for example security, to support the operating system.
Compression	Designed by an external organisation, often as an alternative software solution.

Grade 4–6

3 Describe the process of memory management within an operating system. **[2]**

....

....

....

....

Grade 4–6

4 A new webcam and microphone have been plugged into a home computer.

What will the operating system need to install or update before it can communicate with them? **[1]**

....

System Security and Software

Grade 4–6

5 Why might the following types of organisation need encryption software?

a) School [1]

b) Games design company [1]

c) Social network [1]

d) Hospital [1]

Grade 4–6

6 A small printing business has contacted a computer support consultant because their office machine is running slowly.

The consultant recommends using defragmentation utility software.

What is this and how does it work? [3]

Ethical, Legal, Cultural and Environmental Concerns

Ethical and Legal Concerns

Grade 4–6

1 A government employee loses his smartphone when travelling for work. He needs to retrieve it because he is concerned about its contents.

Describe **three** systems that could be used to try to locate it. **[3]**

_____

_____

_____

Grade 4–6

2 Judy joins a large social network and is surprised to find that all of the services it offers are free.

Explain how large social networks generate income. **[4]**

_____

_____

_____

_____

Grade 4–6

3 Governments around the world would like to monitor our digital footprint so that they can identify potentially dangerous threats.

Provide **three** examples of what the term 'digital footprint' refers to. **[3]**

_____

_____

_____

Grade 4–6

4 Describe **three** examples of situations where automated devices can work in locations potentially dangerous to humans. **[3]**

_____

_____

_____

Ethical, Legal, Cultural and Environmental Concerns

Grade 4–6

5 Describe how criminals are using the Internet to commit offences in the following areas.

a) Films and TV   [1]

_____

b) Banking   [1]

_____

c) Motor vehicles   [1]

_____

d) Illegal substances   [1]

_____

Grade 7–9

6 Ray is sharing a house at university. He receives a letter from his ISP saying he has been breaking copyright laws and his contract may be cancelled. He does not understand why this has happened.

State **two** possible reasons why this has happened.   [2]

_____

_____

Cultural and Environmental Concerns

Grade 1–3

1 Briefly describe the term 'digital divide'.   [2]

_____

_____

Grade 4–6

2 State an example of how computer technology might affect the life of a typical adult in each of the following situations.

a) Accessing local and international news at breakfast time   [1]

_____

b) Listening to music on the train to work   [1]

_____

c) Working as an estate agent [1]

....

d) Booking a holiday over lunch [1]

....

e) Talking to family during the evening [1]

....

f) Evening meal time [1]

....

g) Watching a movie [1]

....

h) Reading [1]

....

3 Explain how online tutorials and videos have changed the ways in which children and adults can learn new skills. [4]

....

....

....

....

Ethical, Legal, Cultural and Environmental Concerns

Grade 4–6

4 Complete the following table on the positive and negatives impacts of technology on the environment. Add a tick in the 'positive' or 'negative' column for each, as appropriate.   **[10]**

	Positive	Negative
Increased energy consumption of digital devices		
Increased greenhouse gas emissions to meet additional power needs		
Reductions in the amount of paper used		
Use of toxic materials in device manufacture		
Increasingly efficient renewable energy production systems		
Recycling waste materials from outdated or unwanted technology		
Downloads require fewer materials than physical media		
The transportation and use of raw and synthetic materials in the production of smart devices		
Impact on travel and commuting due to increased remote working		
Smart devices being able to control their energy usage		

Grade 4–6

5 Briefly describe why the number of unwanted electrical devices is increasing faster than ever before.   **[2]**

....

....

....

Grade 7–9

6 How might technology help a surgeon to perform an operation from a remote location?   **[2]**

....

....

....

Ethical, Legal, Cultural and Environmental Concerns

Computer Science Legislation

 1 Match each piece of legislation with its definition. [2]

Data Protection Act 2018	To prevent the hacking and damaging of computer systems.
Computer Misuse Act 1990	To provide creators of media with the right to control how their products are accessed and sold.
Copyright, Designs and Patents Act 1988	To protect the personal information held about individuals within organisations.

 2 What is meant by the term 'copyright'? [2]

....

....

....

....

 3 Prabhat is a designer who wants to add high-quality photos to his promotional website.
State **three** legal ways he might achieve this. [3]

....

....

....

....

Grade 4–6

4 Fiona is opening a cupcake business and has been advised to use open source software to save money.

What does the term 'open source' mean? [2]

....

....

Grade 7–9

5 State **four** principles of the Data Protection Act 2018. [4]

....

....

....

....

Algorithms and Computational Logic

Algorithms and Flowcharts

Grade 4–6

1 An amusement park ride program opens and closes an entry gate until the maximum number of people have got on the ride.

What type of algorithm is this? [1]

....

Grade 4–6

2 A programmer has been asked to create a virtual driving simulator.

How will abstraction help to design a solution? [2]

....

....

Grade 4–6

3 Draw the correct flowchart shape for the following functions: [5]

Start/stop	
Input/output	
Decision	
Process	
Sub-program/routine	

Algorithms and Computational Logic

Grade 4–6

4 Explain the purpose of a 'structure diagram' when tackling a programming problem. **[2]**

Searching and Sorting Algorithms

Grade 4–6

1 Match each type of sort to the correct description. **[2]**

Bubble sort	Each item in an unordered list is examined in turn and compared with the previous items in the list.
Merge sort	Pairs of values in a list are compared to each other and swapped until they are in the correct order.
Insertion sort	Data is repeatedly split into halves until each list contains only one item.

Grade 7–9

2 Why must a data set be ordered when a binary search is carried out? **[3]**

Grade 7–9

3 Carry out a bubble sort on the data set (6,2,4,1,8) to put the values in ascending order. **[5]**

Pseudocode 1

Grade 4–6

1 What is the difference between pseudocode and a programming language such as Python? **[2]**

Algorithms and Computational Logic

Grade 4–6

2 State a pseudocode keyword that might be used in a program to provide a response
a) if a statement is met and **b)** if a statement is not met. **[2]**

Grade 4–6

3 Within an algorithm/program, what is the difference between a variable and a constant? **[3]**

Grade 4–6

4 A music streaming algorithm is being written by a team of developers. Why is the use of comments so important? **[2]**

Grade 7–9

5 Explain the term 'naming convention' and provide **two** common coding examples. **[4]**

Grade 7–9

6 Write a simple algorithm/program that is used to check the weight of luggage being put into the hold of a plane. If the value is greater than 30, then 'Too heavy' is displayed; otherwise, 'OK' is displayed. **[4]**

Algorithms and Computational Logic

Pseudocode 2

Grade 1–3

1 Which type of error is a simple typing mistake or use of the wrong character known as?   **[1]**

....

Grade 4–6

2 Describe the purpose of the following short program.   **[3]**

```
stepOne = input("Please enter your password")
stepTwo = input("Please confirm your password again")
if stepOne == stepTwo then
    print("Access granted")
else
    print("Access denied")
end
```

....

....

....

Grade 7–9

3 What does the following pseudocode do?   **[3]**

```
t = 0
while t <= 50
    print t
    t = t + 1
endwhile
```

....

....

....

Grade 7–9

4 Describe the difference between the operators MOD and DIV.   **[2]**

....

....

....

Algorithms and Computational Logic

Grade 7-9

5 Write a simple algorithm/program that asks for the three dimensions required and then calculates and returns the volume of a room. **[5]**

Grade 7-9

6 Write a short algorithm/program that allows cheaper train tickets to be bought for children under 16 OR adults over 65. A cheap ticket costs £10; a standard ticket costs £20. **[3]**

Boolean Logic

Grade 4-6

1 Draw the standard shapes for an AND, OR and NOT gate. **[3]**

Grade 4–6

2 The NOT gate is sometimes referred to as an inverter. Why is this? **[1]**

....

Grade 4–6

3 Identify **two** mistakes in the logic diagram below. **[2]**

A ——————⊐ C   ....

B ——o◁———— D   ....

Grade 4–6

4 Draw a truth table for a standard NOT gate. **[4]**

Grade 7–9

5 Create a truth table for the following logic circuit. **[5]**

A ——⊐ D
B ——— ⊐ K ▷o— Z

Grade 7–9

6 Draw a logic diagram and a truth table for the following:

A NOT gate connected to the output of an AND gate. **[6]**

Programming Techniques 1

1 Nik's Shack, a mountain bike shop, is creating a database of stock bikes.

Complete the following table by selecting a suitable data type for each field.   **[5]**

Field	Example	Data type
bikeBrand	Peak Buster	
numberofGears	18	
overallWeight	10.4	
colourCode	S	
inStock	Yes	

2 What would the following pieces of pseudocode do?

a)   int("1977")   **[1]**

_____

b)   str(3827)   **[1]**

_____

3 What would be the best data type to use in the following scenarios?

a) The entry of an alphanumeric password   **[1]**

_____

b) The precise weight of a metal   **[1]**

_____

c) A Yes/No question about personal food preferences   **[1]**

_____

d) Asking a user for their age in years   **[1]**

_____

Grade 4–6

4 Explain the term 'concatenation' in relation to programming.   **[2]**

....

....

....

Grade 7–9

5 Write a simple algorithm/program that does the following:

• asks for the first and last name of the user

• asks for the year of birth of the user

• takes the first letter of the first name, the first letter of the last name and the year of birth to generate a username, and returns this on screen.   **[4]**

....

....

....

....

....

....

....

Grade 7–9

6 Consider a text file called 'story.txt'

Write a short algorithm/program to open the file and print the first line.   **[4]**

....

....

....

....

....

....

Programming Techniques 2

Grade 1–3

1 State **three** reasons why a programmer may use sub-programs to improve the efficiency of a program. **[3]**

....

....

....

Grade 1–3

2 Nik's Shack now has a database of mountain bikes in stock, which is called 'Stock'.

bikeID	bikeBrand	numberofGears	overallWeight	colourCode	inStock
0001	Peak Buster	18	10.4	S	Yes
0002	PB Wheelers	21	13.2	R	Yes
0003	Peak Buster	24	11.3	G	No
0004	Team PB	18	12.4	B	Yes
0005	PB Wheelers	24	14.3	Y	Yes

a) How many records does the database have? **[1]**

....

b) How many fields does each record have? **[1]**

....

c) What would be the problem with a single character colour code? **[1]**

....

d) Which field is the primary key? .... **[1]**

Grade 7–9

3 Still looking at Nik's Shack 'Stock' database, write SQL queries in order to:

a) list all of the bikes with the colour code 'G'. **[1]**

....

b) search for all bikes that have more than 20 gears. **[1]**

....

c) search for bikes that have 18 gears AND are in stock. **[1]**

....

Programming Techniques, Programs, Translators and Languages

Programming Techniques 3

1 a) Write a short program to create a one-dimensional array of the race laptimes shown. **[2]**

	0	1	2	3
Laptime	59.4	64.3	74.3	81.9

b) How might this be turned into a two-dimensional array? **[1]**

2 Sub-programs can contain both parameters and arguments.

Describe the difference between these. **[2]**

3 Oscar is writing a weather application and will be using multiple sub-programs.

a) Write a short sub-program to carry out a conversion from Celsius (C) to Fahrenheit (F) when required. **[4]**

Note: F = (C × 1.8) + 32

b) Why is Oscar using a function rather than a procedure? **[1]**

Programming Techniques, Programs, Translators and Languages

Producing Robust Programs

Grade 1–3

1 Testing data is an essential part of the design process. Match each term with its definition. **[3]**

Normal data	Values of the correct data type but that cannot be processed as they are outside pre-determined limits.
Boundary data	Incorrect data of a type that should be rejected by the program or system.
Invalid data	Acceptable, error-free data likely to be input into the program.
Erroneous data	Values at the limit of what a program should be able to accept, minimum and maximum dates for example.

Grade 4–6

2 Daisy is writing a program to be used in a school classroom. A friend has reminded her to consider defensive design when designing a system to be used by children.

a) What is defensive design? **[2]**

b) Describe **three** ways in which this can be applied to a system to be used by children. **[3]**

Grade 4–6

3 Andrew is just starting out as a programmer and a colleague provides him with a list of tips to keep his programs well maintained.

Describe **four** tips for good program maintenance. **[4]**

Grade 4–6

4 When checking through a first version of an address book program, Rachel finds several syntax and logic errors.

What are the differences between these two types of error? **[2]**

....

....

Grade 4–6

5 Logic errors are often more difficult to solve. Why might this be the case? **[2]**

....

....

Grade 7–9

6 Pawel is carrying out a range of tests on a computer-based version of a popular board game.

How does boundary and erroneous data differ to normal data? **[2]**

....

....

Languages, Translators and Integrated Development Environment

Grade 4–6

1 Isaac has been asked to choose a high-level programming language to focus on during a university course.

List at least **four** current high-level programming languages open to him. **[4]**

....

....

....

....

Grade 4–6

2 Many programmers still specialise in using low-level languages.

State **two** reasons why this might be the case. **[2]**

....

....

Programming Techniques, Programs, Translators and Languages

Video Solution   Question 3

Grade 4–6

3 Noah has written a financial application using an integrated development environment, and the code editor function has helped him to identify syntax errors.

Describe **two** ways this might have happened.   **[2]**

....

....

Grade 4–6

4 State **three** ways in which an IDE helps a programmer to spot and rectify logic errors.   **[3]**

....

....

....

Grade 7–9

5 Match each IDE functionality with its description.   **[3]**

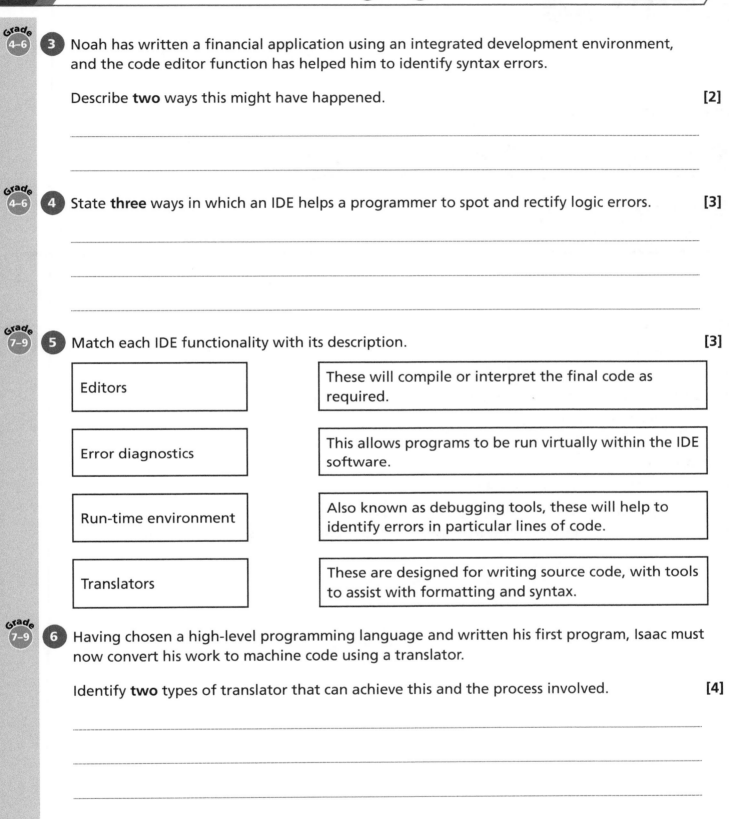

Editors		These will compile or interpret the final code as required.
Error diagnostics		This allows programs to be run virtually within the IDE software.
Run-time environment		Also known as debugging tools, these will help to identify errors in particular lines of code.
Translators		These are designed for writing source code, with tools to assist with formatting and syntax.

Grade 7–9

6 Having chosen a high-level programming language and written his first program, Isaac must now convert his work to machine code using a translator.

Identify **two** types of translator that can achieve this and the process involved.   **[4]**

....

....

....

Collins

GCSE Computer Science
Paper 1: Computer Systems

Time allowed: 1 hour 30 minutes

Instructions

- Use black ink.
- Answer **all** the questions.
- Write your answer to each question in the spaces provided.
- You may **not** use a calculator.

Information

- The total mark for this paper is **80**.
- The marks for questions are shown in brackets **[ ]**.
- Quality of extended responses will be assessed in this paper in questions marked with an *.

Name: ....

1 Elizabeth is a freelance journalist with her own blog. She posts articles, images and videos from all over the UK, but finds that her smartphone and digital camera quickly run out of storage space.

(a) One of the new cameras she is considering has an option to upload images to the 'cloud'.

Explain what is meant by 'cloud storage'.

....

.... **[1]**

(b) Uploading files to the cloud brings advantages and disadvantages.

(i) Describe **two** advantages of storing files online.

....

.... **[2]**

(ii) Describe **two** disadvantages of storing files online.

....

.... **[2]**

(c) Elizabeth is considering upgrading the primary and secondary storage on her desktop computer.

Describe the difference between 'primary storage' and 'secondary storage' and provide an example of each.

....

....

....

.... **[4]**

2   Javed has a gaming PC that he built himself. In recent months, he is finding that the PC has become slower and new games are not running well. His first thought is to check the operating system is up to date.

(a) State **two** reasons why operating systems are regularly updated.

....

....   **[2]**

(b) Javed's friend recommends that he upgrade the RAM in his computer.

(i) Define the term 'RAM'.

....

....   **[1]**

(ii) Explain why having more RAM should improve the performance of a computer.

....

....   **[2]**

(c) Having done some research, Javed decides to try to reconfigure his operating system before he upgrades the hardware. He experiments with virtual memory settings and disk defragmentation.

(i) Define the term 'virtual memory'.

....

....   **[1]**

(ii) State **one** advantage and **one** disadvantage of using virtual memory.

....

....   **[2]**

(iii) Explain the process of disk defragmentation.

....

....   **[2]**

3 A cycle-hire business, specialising in electric bikes, has expanded to a second shop in a woodland area. The business owners make use of computers and tablet devices for bookings, and they need to make sure that they have a reliable system in place.

(a) Devices in the shop are connected to a LAN.

(i) Define the term 'LAN'.

....

.... **[1]**

(ii) The business connects its LAN to a second LAN in the other shop. State the device needed to connect them together, and what could be created as a result?

....

.... **[2]**

(b) Wireless technology is used in the shop and the owners are concerned about its security and which encryption option to use.

(i) State **three** common wireless encryption standards.

....

....

.... **[3]**

(ii) Explain which standard they should use and why.

....

.... **[2]**

(c) The shop also has a children's seating area with tablet computers for children to play cycle games and look at videos while parents make bookings. These devices were designed to meet modern hardware and software standards.

Why are hardware and software standards important for modern devices?

....

.... **[2]**

4   An independent film production company specialises in creating film shorts on a budget that look like more expensive productions. All of their films are shot and edited digitally, and the company is constantly looking for new software and media clips it can use.

(a) State which legislative Act applies to the following aspects of their business.

(i) Concerns about their films being pirated and shared online before official release.

_____ [1]

(ii) Making sure the personal details of all employees and customers are secure.

_____ [1]

(b) When editing film and music tracks, the company can save their files in either lossy or lossless format.

(i) Explain the difference between the two formats.

_____

_____ [2]

(ii) State which format would be the more appropriate when editing content and explain why this would be the case.

_____

_____ [2]

(c) Live bands are often brought in to the editing studio to create a high-quality original recording.

State the name of the process of transferring analogue to digital audio and explain the importance of sample rate and bit depth.

_____

_____

_____

_____ [3]

5   Ian runs a small computer consultancy company, offering advice on infrastructure, networks and security.

   (a) Two of the services that Ian offers are penetration testing and advice on user access levels.

   (i) Define the term 'penetration testing'.

   [1]

   (ii) Define the term 'user access levels'.

   [1]

   (b) Many of the organisations he works with dispose of unwanted, but still serviceable, equipment straight to landfill.

   (i) State **three** reasons for **not** doing this.

   [3]

   (ii) State an ethical alternative to disposing of unwanted equipment.

   [1]

6   A world news organisation is expanding its website to include a subscription service that will pay for additional journalists around the world. The organisation is very concerned with security, as an older version of its website was often attacked.

   (a) The original website was the victim of a DoS attack. Define the term 'DoS attack'.

   [1]

(b) Subscription customers are reminded to create a strong password when setting up their account.

State **three** pieces of password advice.

....

....

.... **[3]**

(c) The subscription page of the site is a HTTPS page.

(i) Define the term 'HTTPS'.

....

.... **[1]**

(ii) HTTPS is a network protocol. Explain what is meant by a network protocol.

....

.... **[1]**

(d) Customers of the original website were often targeted by malware.

State **three** different pieces of malware and explain the damage each can cause.

....

....

....

....

....

....

....

....

....

....

....

....

.... **[6]**

7   Alexander is setting up a client–server network. It will serve multiple workstations in a computer-based training room. He has narrowed it down to three choices; see **Fig. 1.**

Fig. 1

Server 1	Server 2	Server 3
CPU Clock Speed: 3.2 GHz	CPU Clock Speed: 2.4 GHz	CPU Clock Speed: 2.8 GHz
CPU Cores: 2	CPU Cores: 1	CPU Cores: 4
Hard Drive Space: 500 GB	Hard Drive Space: 750 GB	Hard Drive Space: 1 TB

(a) Identify the most appropriate server and give **two** reasons for your choice.

....

....

.... **[3]**

(b) A star network topology has been chosen.

Describe **two** advantages and **two** disadvantages of using this network topology.

....

....

....

.... **[4]**

(c) Explain **three** benefits a network manager will get from using client–server-linked workstations rather than individual computers.

....

....

....

....

....

....

....

.... **[3]**

8   A university has designed an e-learning platform that can be installed on all devices across the campus.

(a) Students have requested that all messages are encrypted.

Define the term 'encryption'.

....

.... **[2]**

(b) Students are also concerned about social engineering.

(i) Define the term 'social engineering'.

....

.... **[1]**

(ii) Describe **three** examples of how it can take place.

....

....

.... **[3]**

9* An international bank is improving its network and security systems across the world. This is in response to concerns about both the online and physical security of the financial data it holds.

Many people are concerned that it is impossible to protect data from theft.

Discuss this statement, considering potential solutions.

....

....

....

....

....

....

....

....

....

.... [8]

Collins

GCSE Computer Science
Paper 2: Computational Thinking, Algorithms and Programming

Time allowed: 1 hour 30 minutes

Instructions

- Use black ink.
- Answer **all** the questions.
- Write your answer to each question in the spaces provided.
- You may **not** use a calculator.

Information

- The total mark for this paper is **80**.
- The marks for questions are shown in brackets **[ ]**.

Name: ....

SECTION A

1. Finn manages an electric car showroom and has a database of regular customers. The database is called VipCustomers; see **Fig. 1**.

Fig. 1

custID	surname	firstName	carsBought	houseNumber	postCode	contactNumber
0001	Peak	Ray	3	161	VC1 4RD	07123827645
0002	Peak	Judy	2	9	VC2 7YT	01293837645
0003	Ibex	Rob	4	32	VC7 3EK	01293695641
0004	Ibex	Katie	1	4	VC2 4RD	07256453726

(a) Describe the difference between a record and a field.

....

.... **[2]**

(b) Create database searches using Structured Query Language (SQL) to display the following.

(i) All the records of customers who have bought only one car.

....

.... **[1]**

(ii) The first name of customers who have bought three or more cars.

....

.... **[1]**

(iii) The first name, surname and contact number of customers from the VC2 4RD postcode.

....

....

.... **[1]**

2   Consider the following data sequence: 13, 32, 10, 19

(a) Show the stages of a bubble sort when applied to this sequence.

....

....

....

.... **[3]**

(b) Describe the process of carrying out a linear search on the same sequence to find the value 10.

....

....

.... **[3]**

(c) A linear search is inefficient for large datasets.

State the type of search that should be used for large datasets.

.... **[1]**

3   Catherine manages a netball team and is building a program to record the goals scored. **Fig. 2**, titled goalsScored, shows the player numbers of three team members and the number of goals they have scored.

Fig. 2

	0	1	2	3
0	12	3	3	2
1	13	2	4	5
2	14	6	4	7

(a) It has been recommended that Catherine should use an array.

(i) Write an algorithm to create this table as a two-dimensional array.

[3]

(ii) Write a short search algorithm to return the number of goals scored by player number 13 in their third game.

[1]

(b) In respect to the amount of data it can store, state how a one-dimensional array differs from a two-dimensional array.

[1]

4   Consider a two-level logic circuit with three inputs, A, B and C, and an output, X.

(a) Create a logic circuit diagram that represents the Boolean expression:

(i) X = (A OR B) AND (NOT C).

[3]

(ii) Complete the following truth table for the same expression.

A	B	C	X

[8]

(b) Describe the purpose of the following logic circuit.

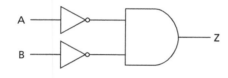

....

.... [2]

5   Noah is part of a development team that writes smartwatch applications. He specialises in small graphical and audio elements. A graphical icon has the hexadecimal code: 69 96

(a) Convert the hexadecimal code into a series of binary numbers and complete the table below to show the icon. Use 0 = White and 1 = Black.

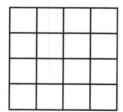

[5]

(b) The application is limited to a 4-bit image.

State how many colours this would allow.

.... [1]

(c) File compression is used to save all images used in the application.

Describe **one** advantage and **one** disadvantage of this method.

_____

_____

_____

_____ **[2]**

6 **(a)** Insert the missing unit in this ordered list of data storage sizes.

MB   GB   __   PB   **[1]**

(b) Convert the decimal number 249 into an 8-bit binary number.

_____ **[1]**

(c) Trying to convert the number 259 into an 8-bit binary number results in an overflow.

Define the term 'overflow' in this context.

_____

_____ **[1]**

(d) Add the following binary numbers together and give the answer as a binary and a decimal number:

10010010

00001100

_____ **[2]**

(e) Convert the hexadecimal number E9 into a decimal number and show your working.

_____

_____ **[2]**

7   A UK mobile telephone number in the format 07######## will often need to be written in the international format +447########.

(a) Write an algorithm that asks for a UK mobile number, replaces the first 0 with the prefix +44 and returns the result to the user. It should also return 'not recognised' for numbers not starting with a '0'.

[5]

(b) The international UK number begins with the + symbol.

State the most suitable data type for handling this symbol.

[1]

SECTION B

Some questions require you to respond using either an exam-style reference language or a high-level programming language. These will be clearly labelled.

8   Emelia is developing a range of small applications to be built into a child-friendly tablet computer.

(a) Her first program (see **Fig. 3**) simulates two dice being rolled to start a board game. The program makes use of the random function and the numbers on both dice need to match for the game to start. If not, they are rolled again.

Fig. 3

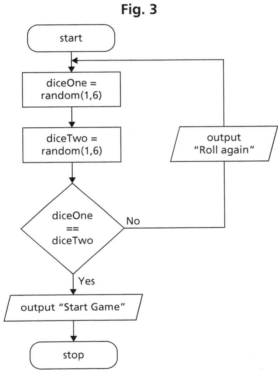

(i) State **two** variables in the program.

....

.... **[2]**

(ii) Explain the process taking place within the diamond shape.

....

.... **[2]**

(b) In order to process dice roll outcomes, the data needs to be stored.

What data type would be the most appropriate for the results of a dice roll?
Tick **one** box and explain your choice.

Data Type	Tick one box
String	
Integer	
Real	
Boolean	

Explanation: ....

.... **[2]**

(c) Emelia would now like to write this as a sub-program.

(i) Describe whether this program would be a function or a procedure and explain
your choice.

....

.... **[2]**

(ii) Write a short program representing the flowchart as a sub-program.
*You must use **either**:*

- *an exam-style reference language* ***or***

- *a high-level programming language you are studying.*

_____

_____

_____

_____

_____

_____

_____

_____

_____

_____

[4]

(d) Complete the test plan below for the flowchart in **Fig 3**.

Test Data	Test Type	Expected Result
diceOne = 2 diceTwo = 3	Normal	Roll again
diceOne = 1 diceTwo = 6	Boundary	
diceOne = 9 diceTwo = 7		
diceOne = 4 diceTwo = 4		

[5]

(e) Emelia's next program is a child-friendly weather application. Part of the program converts Celsius into Fahrenheit. Users are asked to enter a temperature in Celsius; a calculation is carried out and the value is returned to them in Fahrenheit.

```
tempC = input("Please enter the temperature in Celsius")
tempF = tempC * 1.8
tempF = tempF + 32
print (tempF)
```

(i) Complete the trace table below using an input value of 10 degrees Celsius.

Line	Input (TempC)	tempF	Output
1			
2			
3			
4			

[4]

(ii) The program is being refined to also display whether the current temperature in Fahrenheit is above or below the average temperature for the time of year. The average is represented by the variable 'averageTemp'. Improve the original program to include this functionality.

*You must use **either**:*

- *an exam-style reference language **or***

- *a high-level programming language you are studying.*

[4]

(f) Emelia has also been asked to add a feature that checks how long the user has been using the tablet. It displays a warning if it is over the three-hour limit on any one day.

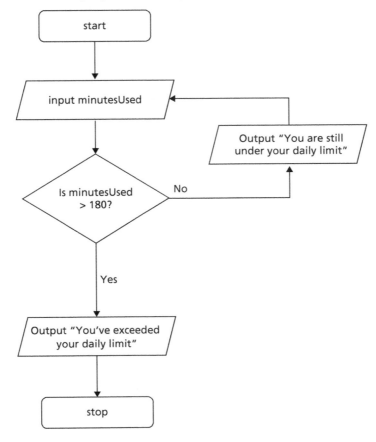

Rewrite the flowchart as a program.

*You must use **either**:*

- *an exam-style reference language **or***
- *a high-level programming language you are studying.*

....

....

....

....

....

....

[4]

....

Notes

Answers

The Purpose and Function of the Central Processing Unit

1.

Input	A device that sends information or data to the CPU.
Output	A device that receives instructions or data from the CPU.
Software	Programmed instructions for a computer with a specific task.
Hardware	Any physical device that would normally form part of a computer system.

[3]

2.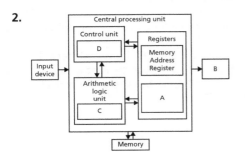

[3]

3. Memory [1]
4. Stored program means that both the computer program and the data it processes are stored in memory [2]
5. The arithmetic logic unit (ALU) is where calculations and logic comparisons are carried out, and the results of these calculations and logic comparisons are stored in the accumulator until they are needed [2]
6. The control unit [1]

Systems Architecture

1. Clock speed is the rate at which instructions are processed by the CPU per second [1]
2. The whole system may still function if a non-essential embedded system is damaged [1]. Different manufacturers can simultaneously work on embedded systems during production [1]
3. Multicore processors have more than one CPU on the same chip [1]; this means that tasks can be carried out simultaneously, which speeds up the system [1]
4. Any **four** of the following:
   An instruction is fetched from memory [1]
   The instruction is then decoded [1]
   The decoded instruction is then executed so that the CPU performs continuously [1]
   The process is repeated [1]
   The program counter is incremented [1]
   The instruction is transferred to the MDR [1]
   The address of the instruction to be fetched is placed in the MAR [1]

5. Accept either of the following:
   L1 cache is smaller [1], but faster [1].
   L2 cache is larger [1], but slower [1].
6. Embedded systems are often installed deep into the machine [1] and if they fail can be difficult to replace or repair [1].

Memory

1. RAM is a temporary area that a computer uses to store data in current use [1]
2. a) Volatile means that once power is switched off, all stored data is lost [1]. Non-volatile means that any instructions written are permanently kept without power [1].
   b) Manufacturers can write instructions to ROM, such as BIOS, and these cannot be changed or edited [1]
3. The main memory component of a computer system [1]
4. True [1]
5. a) Basic input/output system [1]
   b) It ensures hardware communications [1]; it starts running the operating system [1]
6. a) It is created because the RAM becomes full [1]
   b) It is created on the hard drive [1]
   c) As secondary storage communication is not as fast as RAM, the system will become slow if secondary storage communication is used too much [1]

Storage Types, Devices and Characteristics

1. Magnetic storage is large [1] and has complex moving parts that could be damaged with physical use [1]
2. Any five of the following: capacity [1]; speed [1]; portability [1]; durability [1]; reliability [1]; cost [1]
3. a) CD [1]; DVD [1]; Blu-ray [1]
   b) Discs can be damaged easily [1]; capacity is limited by type [1]; the correct writer/player must be used [1]
4. The original device has more storage capacity [1], but the new device will run faster [1]
5. Any **two** of the following: An optical drive is slower to access than others [1], is liable to skip/jump if it is moved [1] and has limited capacity [1]
6. At regular intervals, move the stored files to a new storage media technology [1]

Units and Formats of Data

1.

high-definition video	gigabytes
MP3 audio file	megabytes
system backup files	terabytes
word processing document	kilobytes

[3]

2. a) 156 [1]
   b) 246 [1]
   c) 75 [1]
   d) 33 [1]

Answers

3. Character or coding groups [1]
4. Unicode [1]
5. To provide capacity to represent another 128 characters [1]; to allow it to be used with common 8-bit systems [1]
6. Examples: Backspace [1]; Enter (Carriage return) [1]; Escape [1]; Tab [1]

Converting Data 1
1. a) 00010010 [1]
   b) 00011001 [1]
   c) 11111011 [1]
   d) 10100001 [1]
2. a) 11101010 [1] 234 [1]
   b) 01011001 [1] 89 [1]
3. Example: 11111100 + 10000000 (380) [1]
4. a) 01101010 [1]
   b) 01001100 [1]
5. a) 01101010 [1]
   b) 00101101 [1]
6. To carry out multiplication and division [1]

Converting Data 2
1. a) FE [1]
   b) 63 [1]
   c) 2A [1]
2. 42 49 47 [3]
3. a) 79 [1]
   b) 140 [1]
   c) 18 [1]
4. Because long binary sequences can be shortened to a more manageable hexadecimal sequence [1]
5. They might see it as the denary number 24 rather than the characters 2 and 4 [1]
6. It is a shortcut reference [1] language, created by programmers [1]

Audio/Visual Formats and Compression
1. a) 1 bit [1]
   b) 2 bits [1]
   c) 4 bits [1]
2. To catalogue data by (any two of): location [1]; date [1]; time [1]; camera settings [1]
3. RAW is a lossless file format, so all the original image data is maintained [1]
4. MP3 [1], because the small file size is easy to download [1], and the quality can be adjusted to suit user requirements [1]
5. Benefit: high-quality accurate recording [1]; drawback: large digital file size [1]
6. The file type JPEG uses lossy compression [1], so each time the file is saved, more data is lost [1]

Pages 15–19 Computer Networking

Wired and Wireless Networks 1
1. Local area network (LAN) [1]; wide area network (WAN) [1]. A WAN is formed by connecting two or more LANs together [1] across large distances [1]
2. Fibre-optic cables [1]; satellites [1]
3. Smart TV [1]; smartphone [1]; Internet radio [1]; games console [1]; tablet [1]; media streamer [1]

4. Network bandwidth [1]; interference from external factors and devices [1]; the number of users connecting at the same time [1]
5. He may have not checked the permission settings of the network [1] and so other users have been allowed access to his files [1]
6. Advantages: software and security settings are controlled centrally [1]; client computers can be of relatively low specification [1]. Disadvantages: if the server fails, so does the network [1]; low-specification client machines can run quite slowly [1]

Wired and Wireless Networks 2
1. a) Files can be saved and accessed anywhere [1]; word processing and image editing software can be accessed via a browser [1]; storage devices do not need to be carried [1]
   b) Without an Internet connection, cloud services cannot be accessed [1]
2. A switch connects network compatible devices together on the same network [1], whereas a router connects different networks together [1]
3. Shortest first: Wi-Fi, Ethernet, fibre optic [2]
4. A system to publish linked pages written in HTML [1] that can be viewed using a web browser anywhere in the world [1]
5. Domain Name Service (or Servers) [1] is an Internet naming service that links the IP address of a computer on a network to a text-based website address that is easier to remember [1]

Network Topologies
1.

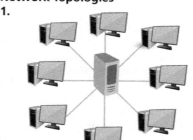

Star Topology

Mesh Topology [2]

2. Any of the following:
   Star advantages: the failure of a device, as long as it is not the server, will not halt the network [1]; the network can be expanded by adding devices [1]; localised problems can be identified quickly [1]; data can be directed to a specific address via the central server [1]

Answers

Star disadvantages: if the server fails, the whole network will collapse **[1]**; extensive cabling and technical knowledge is needed to maintain the server **[1]**

Mesh advantages: all devices share the network load **[1]**; if a device fails, the network will continue to run **[1]**; adding more devices will not affect the speed of the network **[1]**

Mesh disadvantages: managing the network requires a high level of network expertise **[1]**; it can be expensive to set up because of the number of devices required **[1]**

3. The server **[1]**. It directs the flow of data between devices **[1]**.
4. Every device in the network needs to be connected to every other device in the network **[1]**
5. Mesh **[1]**

Protocols and Layers
1. Gigabits per second **[1]**
2. a) An open unsecure network can be connected to by any device **[1]**; these devices may pass malware onto the network **[1]** and can potentially steal personal information **[1]**
   b) An encrypted network connection **[1]** using WPA, WPA2 or WPA3 **[1]**
3. Elderly couple: Post Office Protocol **[1]**, as all emails are downloaded to their home **[1]**. Florist: IMAP **[1]**, as emails can be synced at home and on mobile devices on the move **[1]**
4. Internet (or Network) layer **[1]**
5. Hardware standards allow components from different manufacturers to be connected **[1]** and to be connected to the Internet **[1]**. Software standards allow applications to be installed on common operating systems **[1]** and use common file types **[1]**.

Pages 20–24 System Security and Software

Common System Threats
1. Any four of the following: usernames **[1]**; passwords **[1]**; bank account numbers **[1]**; personal email addresses **[1]**; answers to secret questions **[1]**; full names **[1]**
2.

Pharming	The redirection from a user's website to a fraudulent site, by modifying their DNS entries.
Ransomware	Limits or denies a user access to their system until a ransom is paid.
Trojan	Installed by a user who thinks it is a legitimate piece of software when it is, in fact, malware.

[2]

3. A virus must be transferred from one computer to another via another file **[1]**, for example an email attachment, whereas a worm can replicate itself between systems **[1]**

4. Telephone a member of the public, pretending to be their bank **[1]**, ask them to confirm and obtain their bank details following a fictitious security problem **[1]**, and then use these to commit a crime **[1]**
5. The website was flooded with false data traffic **[1]**, causing the server to crash **[1]**, and this will be repeated until the song is withdrawn **[1]**
6. The transfer of files to and from the workplace **[1]**; Internet browsing rules **[1]**; the use of personal devices in the workplace **[1]**

Threat Prevention
1. As new virus codes appear every day **[1]**, anti-malware software must be updated to include the latest patches **[1]**
2. Any five of the following: make sure they are at least eight characters long **[1]**; use upper- and lower-case characters **[1]**; include special characters **[1]**; avoid real dictionary words **[1]**; avoid any personal information **[1]**; regularly change any password **[1]**
3. Encrypted files can be stored safely **[1]**, with no external access from unwanted users **[1]**, and intercepted messages cannot be read **[1]**
4. Ensuring that staff cannot access personal information **[1]**; making sure that any sensitive data cannot be removed from the network **[1]**; making sure that external devices, which potentially carry viruses, cannot be used **[1]**
5. Any three of the following: weak passwords **[1]**; previously unknown access methods **[1]**; system areas vulnerable to virus attack **[1]**; potential SQL injection areas **[1]**
6. A public key is known by all and is a method used to encrypt a message **[1]**, but the private key needed to decrypt the message is known only to the recipient of the message **[1]**.

System Software
1. Users do not have to use command prompt text functions **[1]** and users can visually drag and drop files **[1]**.
2.

Third-party applications	Designed by an external organisation, often as an alternative software solution.
Utility software	Performs specific tasks, for example security, to support the operating system.
Operating system	The software link between the hardware, software and user.
Compression	The process of reducing the file size of a computer file to use less disk space.

[3]

Answers

3. The managing and allocating of free space and prioritising the amount of memory [1] and resources that the CPU and memory modules can use [1]
4. Device drivers [1]
5. a) To keep students' personal details private [1]
   b) To protect unreleased games from being accessed and stolen [1]
   c) To keep the usernames and passwords secure [1]
   d) To protect the medical records of patients [1]
6. Defragmentation utility software analyses data and how it is stored on a disk [1]. It then rearranges files into a more logical sequence [1] to allow faster access [1]

Pages 25–30 Ethical, Legal, Cultural and Environmental Concerns

Ethical and Legal Concerns

1. Any three from: Triangulation using the mobile phone network [1]; GPS [1]; connection to Wi-Fi networks [1]; operating system or service provider 'lost phone' applications [1]
2. Organisations pay the social network to place advertising on the network [1]. Users' browsing habits on the network are tracked [1] and advertising is targeted at them [1] based on what they like and what their friends like [1]
3. Any three of the following: records of the websites we visit [1], the contents of instant messages/emails [1], the people we communicate with [1] and the locations of the devices we use [1]
4. Examples: scientific studies of a volcano [1]; bomb disposal [1]; deep underwater or space exploration [1]; nuclear or related power generation [1]
5. a) Sharing illegally obtained copies of new films/TV shows online [1]
   b) Stealing login details and transferring money to their own accounts [1]
   c) Selling cars through online auction sites without official paperwork [1]
   d) Selling drugs via online auction/dark websites without medical knowledge [1]
6. Examples: Someone in his household may have done two of the following: used a peer-to-peer network to download films/TV shows [1]; downloaded MP3s from an unofficial website [1]; shared with friends links to websites illegally offering films/TV shows [1]; or someone nearby may be accessing their Wi-Fi without permission [1]

Cultural and Environmental Concerns

1. The digital divide is the social and economic gap [1] between those who have and those who do not have access to computer technology [1]
2. Examples:
   a) Watching news on a laptop via streaming site or listening to news via a smartphone radio app [1]
   b) Using a music streaming app/using wireless headphones [1]
   c) Placing advertisements around the world/receiving photos and text via email [1]
   d) Browsing destinations/watching video reviews/using comparison sites [1]

   e) Using video chat to talk to relatives [1]
   f) Ordering a takeaway online or using an online recipe book [1]
   g) Downloading or streaming a film/booking cinema tickets online [1]
   h) Using an e-reader or reading a physical book that was ordered online [1]
3. Examples: Learning material can be accessed at any time [1] and in any place with an Internet connection [1]; material can be accessed that covers just about any topic [1] and it can be followed at the learners' own pace [1]
4.

	Positive	Negative
Increased energy consumption of digital devices		✓
Increased greenhouse gas emissions to meet additional power needs		✓
Reductions in the amount of paper used	✓	
Use of toxic materials in device manufacture		✓
Increasingly efficient renewable energy production systems	✓	
Recycling waste materials from outdated or unwanted technology	✓	
Downloads require fewer materials than physical media	✓	
The transportation and use of raw and synthetic materials in the production of smart devices		✓
Impact on travel and commuting due to increased remote working	✓	
Smart devices being able to control their energy usage	✓	

[10]

5. Users want the latest technology [1] before current technology comes to the end of its natural life [1]
6. Robotic/virtual technology could mimic the movements of the surgeon [1] across the Internet and recreate it at the patient's location [1]

Answers

Computer Science Legislation

1.

Data Protection Act 2018	To protect the personal information held about individuals within organisations.
Computer Misuse Act 1990	To prevent the hacking and damaging of computer systems.
Copyright, Designs and Patents Act 1988	To provide creators of media with the right to control how their products are accessed and sold.

[2]

2. Copyright is the legal right of the creators of music, books, films and games **[1]** to control how their products are accessed and sold **[1]**

3. Any three of the following: Take the photos himself **[1]**; purchase royalty-free images online **[1]**; download photos from a website that provides free images for commercial use **[1]**; seek permission to use copyrighted material **[1]**

4. Open source software can be freely downloaded and shared online **[1]**, with no limitations on its use **[1]**

5. Any four of the following:
- Data should be used fairly, lawfully and transparently. **[1]**
- Data must be obtained and used only for specified purposes. **[1]**
- Data shall be adequate, relevant and not excessive. **[1]**
- Data should be accurate and kept up to date. **[1]**
- Data should not be kept for longer than necessary. **[1]**
- Data must be kept safe and secure. **[1]**
- Those organisations working with our data are accountable for all data protection and must produce evidence of their compliance. **[1]**

Pages 31–36 Algorithms and Computational Logic

Algorithms and Flowcharts

1. Iteration **[1]**

2. Abstraction is the removal of unnecessary information **[1]**; focusing on the car and the road rather than on the surroundings will help to create a solution **[1]**

3.

Start/stop		**[1]**
Input/output		**[1]**
Decision		**[1]**
Process		**[1]**
Sub-program / routine		**[1]**

4. A structure diagram is used to break down a problem into smaller problems **[1]**; each problem can also be broken down into sub-levels **[1]**

Searching and Sorting Algorithms

1.

Bubble sort	Pairs of values in a list are compared to each other and swapped until they are in the correct order.
Merge sort	Data is repeatedly split into halves until each list contains only one item.
Insertion sort	Each item in an unordered list is examined in turn and compared with the previous items in the list.

[2]

2. A binary search starts at the middle value **[1]**, then splits the data set **[1]** according to whether the data sought is above or below the middle value **[1]**

3. Five steps:
(6,2,4,1,8) to (2,6,4,1,8)	**[1]**
(2,6,4,1,8) to (2,4,6,1,8)	**[1]**
(2,4,6,1,8) to (2,4,1,6,8)	**[1]**
(2,4,1,6,8) to (2,1,4,6,8)	**[1]**
(2,1,4,6,8) to (1,2,4,6,8)	**[1]**

Pseudocode 1

1. Pseudocode is not a real programming language designed to run on a computer **[1]**, so mistakes and plain English terms do not prevent it from being understood **[1]**

2. a) then **[1]**; b) else **[1]**

3. The value stored in a variable can be changed while the program is running **[1]** but a constant cannot **[1]**. A constant value is assigned in the program code **[1]**.

Answers

4. Without comments, it might be difficult for the developers to understand the reasons for each other's coding choices **[1]**; the developers can leave messages within the code to help each other **[1]**
5. A naming convention refers to the naming of variables with a simple rule **[1]** and keeping that rule applied throughout all similar programs **[1]**
Examples: using two words to define a variable but removing the space and using a capital letter on the second **[1]**, for example engineSize or fuelTank **[1]**
6. Example:

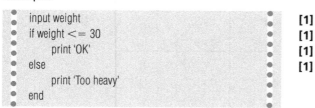

input weight	**[1]**
if weight <= 30	**[1]**
print 'OK'	**[1]**
else	**[1]**
print 'Too heavy'	
end	

Pseudocode 2

1. Syntax error **[1]**
2. A password is asked for twice **[1]**: if the two passwords match exactly **[1]** then access is granted, and if they do not match then access is not granted **[1]**
3. Counts **[1]** and prints **[1]** numbers from zero up to and including 50 **[1]**
4. MOD returns the remainder after a division **[1]**, while DIV divides but returns only a whole number (also known as an integer) **[1]**
5. Example:

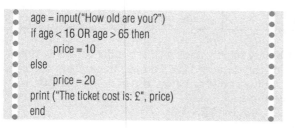

length = input("what is the length of the room?")	**[1]**
width = input("what is the width of the room?")	**[1]**
height = input("what is the height of the room?")	**[1]**
roomVol = length * width * height	**[1]**
print roomVol	**[1]**
end	

6. Example:

age = input("How old are you?")	**[1]**
if age < 16 OR age > 65 then	
price = 10	**[1]**
else	
price = 20	
print ("The ticket cost is: £", price)	**[1]**
end	

Boolean Logic

1. AND Gate

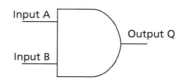

OR Gate

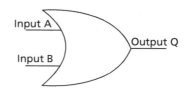

NOT Gate

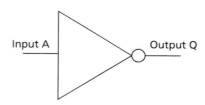

[3]

2. The NOT gate inverts the signal, so a 0 input becomes a 1 output and vice versa **[1]**
3. The NOT gate is reversed **[1]** and the AND gate has two outputs instead of one **[1]**
4.

Input	Output
A	Q
0	1
1	0

[4]

5.

A	B	D	K	Z	
0	0	0	0	1	**[1]**
0	1	1	1	0	**[1]**
1	0	1	0	1	**[1]**
1	1	1	1	0	**[1]**

[1]

6.

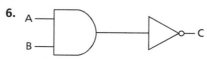

[2]

Inputs		Output
A	B	C
0	0	1
0	1	1
1	0	1
1	1	0

[4]

Answers

Programming Techniques 1

1.

Field	Example	Datatype
bikeBrand	Peak Buster	String
numberofGears	18	Integer
overallWeight	10.4	Real
colourCode	S	Character
inStock	Yes	Boolean

[5]

2. a) This converts the string 1977 to an integer **[1]**
   b) This converts the number 3827 to a string **[1]**
3. a) String **[1]**
   b) Real **[1]**
   c) Boolean **[1]**
   d) Integer **[1]**
4. Concatenation is the process of connecting together two or more strings **[1]**, linking them together as one **[1]**
5. Example:

```
string1 = input("What is your first name?")          [1]
string2 = input("What is your last name?")
string3 = input("What year were you born?")
userName = string1,[0] + string2,[0] + string3       [2]
    print(userName)                                  [1]
end
```

6. Example:

```
myFile = openRead("story.txt")    [1]
lineOne = myFile.readLine()       [1]
print(lineOne)                    [1]
myFile.close()                    [1]
```

Programming Techniques 2

1. To save time **[1]**, to avoid repetitive code **[1]** and to organise/structure programs **[1]**
2. a) 5 **[1]**
   b) 6 **[1]**
   c) More than one colour could begin with the same letter **[1]**
   d) bikeID **[1]**
3. a) SELECT bikeID FROM Stock WHERE colourCode = "G"; **[1]**
   b) SELECT bikeID FROM Stock WHERE numberofGears > 20; **[1]**
   c) SELECT bikeID FROM Stock WHERE numberofGears = 18 AND inStock = "Yes"; **[1]**

Programming Techniques 3

1. a) Example:

```
array laptime[4]       [1]
laptime[0] = "59.4"    [1]
laptime[1] = "64.3"
laptime[2] = "74.3"
laptime[3] = "81.9"
```

   b) By adding the times of several runners **[1]**

2. Parameters refer to variables within a sub-program **[1]**, and arguments are the actual data passed to the parameters **[1]**
3. a) Example:

```
function conversion()                             [1]
    cTemp = input("Enter the temperature in C")   [1]
    fTemp = cTemp * 1.8 + 32                       [1]
return(fTemp)                                     [1]
endfunction
```

   b) A function will return a value; a procedure will not **[1]**

Producing Robust Programs

1.

Normal data	Acceptable, error-free data likely to be input into the program.
Boundary data	Values at the limit of what a program should be able to accept.
Invalid data	Values of the correct data type but that cannot be processed as they are outside pre-determined limits.
Erroneous data	Incorrect data of a type that should be rejected by the program or system.

[3]

2. a) Making sure to consider all those who will be using your program **[1]** and what access each user will be given **[1]**
   b) Examples: use of usernames and passwords **[1]**; making sure children have access to relevant areas only **[1]**; considering what might happen when incorrect keys are pressed **[1]**
3. Use of comments within the program **[1]**; use of indentation **[1]**; use of well-named variables **[1]**; use of sub-programs **[1]**
4. Syntax errors break the rules of the language and will prevent a program from running **[1]**; a program with a logic error will run but provide an unexpected result **[1]**
5. A logic error may be written correctly without any syntax errors **[1]** but references between lines of code may be confused and can often only be spotted by working through the program one line at a time **[1]**.
6. Boundary data tests the ranges of expected values that may be entered **[1]** and erroneous data considers incorrect entries that should not be processed **[1]**.

Languages, Translators and Integrated Development Environment

1. Examples: Python **[1]**; C Family **[1]**; Java **[1]**; JavaScript **[1]**; Visual Basic **[1]**; PHP **[1]**; Delphi **[1]**; SQL **[1]**; Bash **[1]**
2. Any two of the following: they may prefer the fine control and lower memory usage **[1]**; they may specifically want to focus on CPU processing **[1]**; they may be working with older devices **[1]**
3. Code is colour coded, visually highlighting errors **[1]** and language-specific coding mistakes are highlighted when the program is run **[1]**
4. The program can be run virtually before being translated **[1]**; data type errors can be spotted **[1]**; sections of program can be run and checked **[1]**

Answers

5.

Editors	These are designed for writing source code, with tools to assist with formatting and syntax.
Error diagnostics	Also known as debugging tools, these will help to identify errors in particular lines of code.
Run-time environment	This allows programs to be run virtually within the IDE software.
Translators	These will compile or interpret the final code as required.

[3]

6. Compilers [1] convert whole programs to machine code [1]. Interpreters [1] convert one line of code at a time to machine code [1].

Pages 44–53 **Paper 1: Computer Systems**

1. a) The remote storing and accessing of files and applications via the Internet [1]
   b) i) Two advantages from the following: files can be accessed from any Internet-connected location [1]; additional storage devices do not need to be carried [1]; access to files can be shared with other users [1]
      ii) Two disadvantages from the following: loss of access if Internet connection is lost [1]; speed of access is determined by Internet connection [1]; access is not available in all geographical areas [1]
   c) Primary storage describes the main memory of the computer [1]
      Any example from: RAM [1]; ROM [1]; CPU Cache [1]. Secondary storage devices are seperate from the CPU and motherboard [1]
      Any example from: Hard drive [1]; Optical disk (or example of) [1]; SSD (or example of) [1]
2. a) Two reasons from the following: to update security settings [1]; to add functionality [1]; to update drivers [1]
   b) i) Random access memory is a temporary area that a computer uses to store data and instructions in current use [1]
      ii) Additional RAM means more short-term memory to carry out tasks [1], allowing the processor to run and perform better [1]
   c) i) Additional short-term memory space created by the CPU on the hard drive if RAM becomes full [1]
      ii) Advantage: creates additional RAM without replacing or adding hardware [1]. Disadvantage: removes storage space from the hard disk [1] OR access to virtual memory is not as fast as RAM [1]
      iii) Files are moved OR grouped together [1]; empty spaces are grouped together [1]
3. a) i) Local area network. Computers are connected with the ability to share data in a small geographical area [1]

   ii) A router is needed to connect them [1]. A WAN could be created [1].
   b) i) Any three of the following: Wired Equivalent Privacy (WEP) [1]; Wi-Fi Protected Access (WPA) [1]; Wi-Fi Protected Access 2 (WPA2) [1]; Wi-Fi Protected Access 3 (WPA3) [1]
      ii) WPA3 [1], as it is the most recent and most secure [1]
   c) These standards allow components and operating systems from around the world [1] to communicate/function with each other [1]
4. a) i) Copyright, Designs and Patents Act 1988 [1]
      ii) The Data Protection Act 2018 [1]
   b) i) Lossy compression permanently removes data from files [1], whereas lossless uses an algorithm to compress data but then reconstructs it without data loss [1].
      ii) Lossless [1] should be used to preserve all original elements after editing [1]
   c) Analogue audio is converted to a digital format using sampling [1] and the higher the sample rate [1] and bit depth [1], the higher the quality.
5. a) i) The search for vulnerabilities within a system that could be exploited for criminal purposes [1]
      ii) User access levels are used to control the information that a specific user, or groups of users, can access, read or edit [1]
   b) i) Three reasons from the following: landfill is growing around the world [1]; the computers could be used by another user [1]; chemicals and hard-to-recycle elements can damage the environment [1]; creating new products increases greenhouse gases [1]; the transportation of products and waste causes pollution [1]
      ii) One alternative from the following: charities will repurpose machines so that they can be given to those without access [1]; the equipment can be passed on to educational establishments [1]
6. a) To flood a website or network with data traffic so that it is brought to a halt [1]
   b) Three pieces of advice from the following: make sure that passwords are at least eight characters long [1]; use upper- and lower-case characters [1]; include special characters (for example ?, # and %) [1]; avoid using real dictionary words [1]; avoid using any personal information [1]; regularly change passwords [1]; never use the same password for more than one system [1]
   c) i) HTTP Secure encrypts communication between server and client [1]
      ii) A set of rules to allow multiple network devices around the world to communicate [1]
   d) Three from the following: virus [1] – a program hidden within another program or file, designed to cause damage to file systems [1]; worm [1] – a malicious program that acts independently and can replicate itself and spread throughout a system [1]; Trojan [1] – installed by a user who thinks that it is a legitimate piece of software when, in fact, it will cause damage or provide access to criminals [1]; spyware [1] – secretly passes information on to a criminal without your knowledge and is often packaged with free software [1]; adware [1] – displays targeted advertising and redirects search requests without permission [1];

Answers

ransomware [1] – limits or denies a user access to their system until a ransom is paid to unlock it [1]; pharming [1] – the redirecting of a user's website – by modifying their DNS entries – to a fraudulent site without their permission [1]

7. a) Server 3 [1], as it has the highest number of cores [1] and the largest hard drive space [1]

b) Two advantages from the following: the failure of one device, as long as it is not the server, will not affect the rest of the network [1]; the network can be expanded by adding devices until the server capacity is reached [1]; localised problems can be identified quickly [1]; data can be directed to a specific address via the central server, reducing traffic [1]

Two disadvantages from the following: if the server fails, then the whole network will collapse [1]; extensive cabling is required [1]; a high level of technical knowledge is required to maintain the server [1]

c) Software can be installed remotely on client machines by the network manager [1]; low-specification client machines can be added or replaced at low cost as required [1]; security and network access can be controlled from the network's central location [1]

8. a) Data is converted into a meaningless form that cannot be read [1] without the decryption key [1]

b) i) Scams, or similar techniques, designed to steal personal information [1]

ii) Phishing (or variation of) via email or messaging [1]; shouldering – trying to look at a PIN [1]; blagging – trying to con someone face-to-face [1]

9. **Mark Band 3 – High Level (6–8 marks)**
Level of detail:
Thorough knowledge and understanding. Wide range of considerations in relation to question. Response is accurate and detailed. Application of knowledge with evidence/examples related to question. Both sides of discussion carefully considered.

Mark Band 2 – Mid Level (3–5 marks)
Level of detail:
Reasonable knowledge and understanding. Range of considerations in relation to question. Response is generally accurate. Application of knowledge relates to content. Discussion of most areas.

Mark Band 1 – Low Level (1–2 marks)
Level of detail:
Basic knowledge and understanding. Limited consideration in relation to question with basic responses. Limited application of knowledge and basic discussion of content.

Content should include any of the following:
- Physical security
  - Use of locks, safe rooms or obstacles
  - Removable hard drives
  - Increased surveillance
  - Biometric scanners: fingerprints, iris, facial, voice
- Online security
  - Use of firewalls to prevent external access
  - Anti-malware/spyware software
  - User access levels – not allowing access outside of a role
  - Use of strong/regularly changing/non-repeated passwords

- Use of encryption to prevent access to data even if stolen
- Legislation
  - Data Protection Act – ensuring that details are secure and up to date
  - Computer Misuse Act – accessing secure information and sharing or using it to commit crime

Pages 54–65 **Paper 2: Computational Thinking, Algorithms and Programming**

Section A

1. a) A record is a single row or entry of related data in a database [1]; a field is a database category within a record [1]

b) i)
```
SELECT * FROM VipCustomers WHERE
carsBought == 1;
```
[1]

ii)
```
SELECT firstName FROM VipCustomers
WHERE carsBought >= 3;
```
[1]

iii)
```
SELECT firstName, surname, contactNumber
FROM VipCustomers WHERE postCode ==
"VC2 4RD";
```
[1]

2. a) 13, 32, 10, 19 (original) > 13, 10, 32, 19 > 13, 10, 19, 32 > 10, 13, 19, 32 > 10, 13, 19, 32 (final) [3]

b) Check the first item 10 = 13 False, check the second item 10 = 32 False, check the third item 10 = 10 Correct [3]

c) Binary search [1]

3. a) i) Example:
```
goalsScored [3,4]
```
[1]
```
goalsScored = [["12", "3", "3", "2"],["13",
"2","4","5"],["14", "6","4","7"]]
```
[2]

(1 mark for first line; 2 marks for second line)

ii) Example:
```
goalsScored
[1,3]
```
[1]

b) A one-dimensional array can only hold a single list of common elements OR a two-dimensional array can hold more than one list [1]

4. a) i)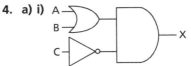

(1 mark for each shape in correct location) [Total 3]

Answers

ii)

A	B	C	X
0	0	0	0
0	0	1	0
0	1	0	1
0	1	1	0
1	0	0	1
1	0	1	0
1	1	0	1
1	1	1	0

(1 mark for each row) **[Total 8]**

b) Both inputs A AND B must be turned off **[1]** to produce a positive output at Z **[1]**

5. a) Binary sequence: 01101001 10010110 **[1]**

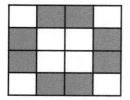

(1 mark for each line) **[Total 5]**

b) 16 **[1]**

c) One each from the following:
Advantage: files can be downloaded quickly **[1]**;
small file sizes mean less storage capacity is used **[1]**
Disadvantage: quality/fine detail can be lost **[1]**;
original image cannot be restored **[1]** **[Total 2]**

6. a) TB **[1]**
b) 11111001 **[1]**
c) 259 doesn't fit into 8 bits, so the computer tries to process more bits than it is designed to handle **[1]**
d) 10011110 **[1]**, 158 **[1]**
e) E = 1110, 9 = 1001, 11101001 = 233 **[2]**

7. a) Example:

```
string1 = input("Please enter your mobile
number")
        if (string1,[0]) = 0
replace (string1,[0]) with (+44) then
        print("Your international number is" +
string1)
else
        print("Number not recognised")
endif
```

**(1 mark – input; 1 mark – string; 1 mark – if;
1 mark – replace number; 1 mark – else)** **[Total 5]**

b) String **[1]**

Section B

8. a) i) diceOne **[1]**, diceTwo **[1]**
    ii) Asking the question **[1]** is diceOne exactly equal to diceTwo? **[1]**

b)

Data Type	Tick one box
String	
Integer	✓ **[1]**
Real	
Boolean	

Explanation: Dice numbers are only whole numbers from 1–6 **[1]**

c) i) Function **[1]** as it returns a calculated value **[1]**
    ii) Example:

```
function diceRoll()
    diceOne = random(1,6)
    diceTwo = random(1,6)
    if diceOne == diceTwo then
        print("Start the game")
    else
        print("Next person")
    endif
    return diceRoll
end function
```

**(1 mark – use of correct variables; 1 mark – use
of procedure; 1 mark – use of random function
correctly; 1 mark – use of print)** **[Total 4]**

d)

Test Data	Test Type	Expected Result
diceOne = 2 diceTwo = 3	Normal	Roll again
diceOne = 1 diceTwo = 6	Boundary	Roll again **[1]**
diceOne = 9 diceTwo = 7	Invalid **[1]**	Roll again **[1]**
diceOne = 4 diceTwo = 4	Normal **[1]**	Start game **[1]**

Answers

e) i)

Line	Input (TempC)	tempF	Output
1	10 [1]	-	-
2	-	18 [1]	-
3	-	50 [1]	-
4	-	-	50 [1]

ii)

```
tempC = input("Please enter the temperature in
Celsius")
tempF = tempC * 1.8
tempC = tempC + 32
if tempF > averageTemp then
      print(tempF + "is above average")
elseif tempF == averageTemp then
      print(tempF + "is average")
elseif tempF < averageTemp then
      print(tempF + "is below average")
else
      print("error")
endif
```

(1 mark – use of correct variables; 1 mark – use of if/else; 1 mark – use of Boolean logic; 1 mark – use of print) **[Total 4]**

f)

```
minutesUsed = input("Input minutes used")
if minutesUsed > 180
      print("You've exceeded your daily limit")
else
      print("You are still under your daily limit")
endif
```

(1 mark – use of correct variables; 1 mark – use of if/else; 1 mark – use of Boolean logic; 1 mark – use of print) **[Total 4]**

Notes